ENERGY

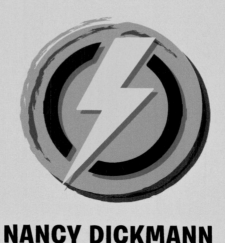

NANCY DICKMANN

BROWN BEAR BOOKS

Published by Brown Bear Books Ltd
4877 N. Circulo Bujia
Tucson, AZ 85718
USA

and

Studio G14, Regent Studios,
1 Thane Villas, London N7 7PH, UK

© 2023 Brown Bear Books Ltd

ISBN 978-1-78121-815-0 (library bound)
ISBN 978-1-78121-821-1 (paperback)

All rights reserved. No part of this book may be reproduced, stored in a retrieval system or transmitted in any form or by any means, electronic, mechanical, photocopying, recording or otherwise, without the prior written permission of the copyright holder.

Library of Congress Cataloging-in-Publication Data available on request

Design: squareandcircus.co.uk
Design Manager: Keith Davis
Children's Publisher: Anne O'Daly

Manufactured in the United States of America
CPSIA compliance information: Batch#AG/5652

Picture Credits
The photographs in this book are used by permission and through the courtesy of:

iStock: srdjanpav 20–21; Shutterstock: aapsky 16–17, Blue Planet Studio 10–11, John Kelly 6–7, Svetlana Lazhko 14–15, scharfsinn 18–19, Sergey Novikov 4–5, CL Shebley 12–13, Pam Walker 8–9.

All other artwork and photography © Brown Bear Books.

t-top, r-right, l-left, c-center, b-bottom

Brown Bear Books has made every attempt to contact the copyright holder. If you have any information about omissions, please contact: licensing@brownbearbooks.co.uk

Websites
The website addresses in this book were valid at the time of going to press. However, it is possible that contents or addresses may change following publication of this book. No responsibility for any such changes can be accepted by the author or the publisher. Readers should be supervised when they access the Internet.

Words in **bold** appear in the Glossary on page 23.

CONTENTS

What Is Energy? 4

Electricity 6

Fossil Fuels 8

Wind and Water 10

Solar Power 12

Heating and Cooking 14

On the Move 16

The Future of Transportation 18

Using Less 20

Quiz 22

Glossary 23

Find out More 24

Index 24

WHAT IS ENERGY?

Energy is the power to do work. Your body uses energy every day. You need energy to breathe and grow. You use energy to dance and swim. We get energy from the food we eat. It's like fuel for the body!

USING ENERGY

We use energy in other ways, too. It takes energy to power cars and trains. Our homes are full of gadgets that use energy. Keeping homes warm takes energy, too. Homes, factories, and vehicles use energy from many different sources.

The more active you are, the more energy you use. You might need a snack in the middle of a match! It will refuel your body.

How We Use Energy

Lots of things in the modern world use energy. Here are just a few of them.

 Lights in homes, offices, and shops run on energy.

Computers, phones, and the internet are powered by energy.

 Tractors, trucks, and airplanes need energy to run.

 Home appliances like washing machines use energy.

 Factories use energy to make the goods we need.

ELECTRICITY

Electricity is a form of energy. It flows in a **current**. When you plug something in, it connects to an electricity supply. Electricity powers lights and fridges. It charges phones and computers. We use it every day.

MAKING ELECTRICITY

You can't dig electricity out of the ground! We have to make it. This happens at a **power plant**. Power plants take other forms of energy. They turn them into electricity. Power plants run day and night. They send electricity to homes and businesses.

Electricity travels through wires. Tall towers carry the wires.

 Burning **fuel** makes heat.

 The heat turns water into steam.

Making Electricity

Most power plants burn fuel. They turn it into electricity. They all work in a similar way.

The steam flows past a **turbine**. Its blades spin.

The turbine connects to a **generator**. Its movement spins the generator.

The generator turns the spinning into electricity.

The electricity is sent out over wires.

Nuclear Power

Nuclear power plants are different. They have fuel rods. The rods do not burn. A special reaction inside them makes heat.

7

FOSSIL FUELS

Oil wells go deep into the ground. They reach pockets of buried fossil fuels.

Power plants need fuel. Many power plants use **fossil fuels**. Crude oil is a fossil fuel. So are coal and natural gas. These fuels burn easily. They release a lot of energy. They are found underground. We dig them out to use them.

RUNNING OUT

We use a lot of fossil fuels. They can only be used once. And our supplies are running low. Fossil fuels take millions of years to form. When they are used up, there will be no more. We need other ways to make electricity.

Different Types

There are different kinds of fossil fuel. We use three main ones.

Coal is a hard rock. It is usually black or brown. It forms in flat layers underground.

Natural gas is a gas. It has no color or smell. It is often found next to crude oil. They are in underground pockets.

Crude oil is a liquid. It comes in different colors. It can be thick or runny.

What's in a Name?

Dinosaur bones are fossils. But fossil fuels aren't made of dinosaurs! They formed from dead plants and animals. These living things died millions of years ago. Over time, their remains turned into fuels.

WIND AND WATER

Wind blows and water flows. They have movement energy. We can turn this movement into electricity. And we don't need fuel to do it! This kind of energy is better for the planet. It will never run out.

HARNESSING THE WIND

Wind turbines are huge. They have long blades. Wind makes the blades spin. This spins gears inside the turbine. They make a generator spin. It produces electricity. A wind turbine only makes electricity when the wind is blowing.

Some wind turbines are on land. Others are at sea. They are grouped together in huge "farms."

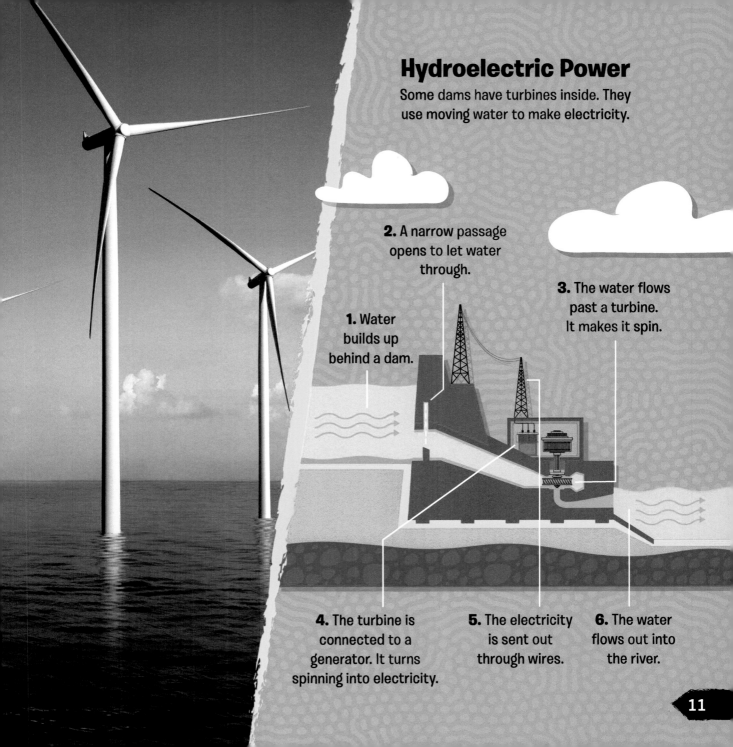

SOLAR POWER

The Sun sends out heat and light. These are both forms of energy. We rely on them. Crops need sunlight and warmth to grow. We can also use the Sun's energy to make electricity. This is called solar power.

TWO KINDS OF POWER

Some solar panels are made up of tiny cells. Each one is a generator. It can turn sunlight into electricity. Other kinds of panels have liquid inside. Sunlight heats the liquid. Its warmth can heat water or inside rooms. Both kinds only work when the Sun shines.

Renewable Energy

Solar power is a kind of renewable energy. So are wind and water power. We will never use up the Sun. We will not run out of wind or moving water. There will always be more!

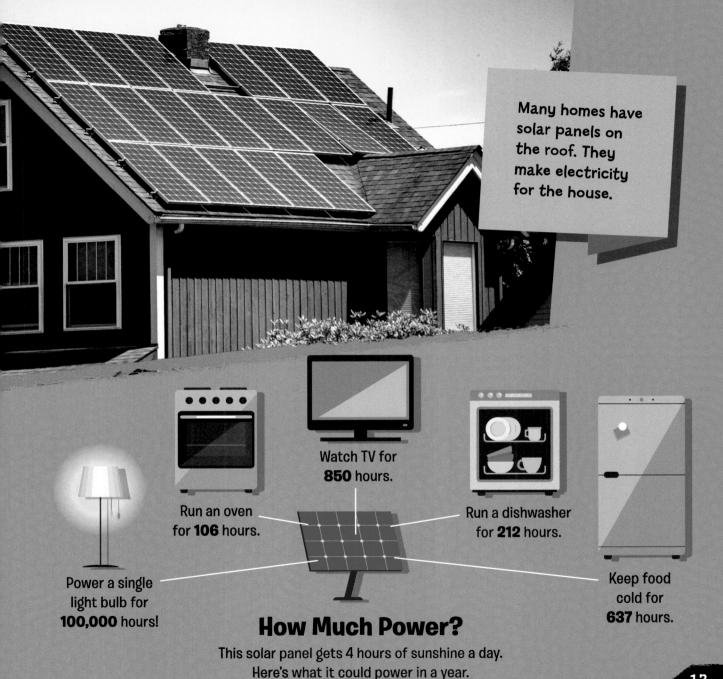

HEATING AND COOKING

Electricity is one form of energy. Heat is another. We need heat to cook our food. We also need it to keep our homes warm. In the past, people used wood fires. They also burned coal. Today, we usually use electricity or natural gas.

NICE AND TOASTY

Some homes have a furnace. It heats air. Then it blows warm air around the house. Other homes have radiators. Hot water flows through them. Both systems can run on electricity. They can run on oil or natural gas.

Biomass Energy

Biomass energy is renewable. We can always grow more plants.

Plants that are burned as fuel are called biomass. The wood in a campfire is a kind of biomass.

Wood fires are good for cooking or keeping warm.

Charcoal is made from wood. It makes heat for cooking.

Corn can be turned into ethanol. This liquid fuel can power cars.

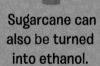

Sugarcane can also be turned into ethanol.

 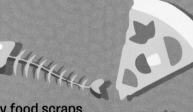

We throw away food scraps and garden waste. This will rot away. It produces gas. We can use the gas as fuel.

ON THE MOVE

How do you get to school? You use energy to walk or bike. Food is your fuel. Do you ride in a bus or car? That needs energy too. Many cars use gasoline. Trucks often use diesel. Both fuels come from crude oil.

BIG VEHICLES

Most cargo ships run on diesel. Airplanes use a fuel called kerosene. It also comes from crude oil. In the past, trains had steam engines. They burned coal to power them. Now many trains use diesel instead.

Most subway trains run on electricity. So do some above-ground trains.

Amazing Oil

Crude oil has its problems. It is still very useful. Here are some of the things we make from it.

 Gasoline for cars and motorcycles.

 Diesel for trucks and ships.

 Kerosene for jet fuel.

 Oils for lubricating engines.

 Wax for candles and crayons.

 Asphalt for patching roads.

 Plastic for toys and clothing.

 Chemicals for cosmetics.

 Propane for camping gas.

What's the Problem?

Most vehicle fuels come from crude oil. Burning them releases **carbon dioxide**. This is a gas. It stays in the **atmosphere**. It traps the Sun's heat. This trapped heat is causing **climate change**.

THE FUTURE OF TRANSPORTATION

Using oil has many problems. We need a cleaner way of getting around. Electricity is a good choice. Electric cars don't give off carbon dioxide. They are quiet and cheap to run. Many people are switching to electric. There are electric motorcycles and scooters, too.

CAN WE ALL GO ELECTRIC?

Scientists are working on other electric vehicles. We already use electric trains. Some companies use electric trucks. Soon we may be using electric cargo ships. We might fly in electric planes.

> You don't put fuel into an electric car. It has batteries. You plug it in to charge them.

Using Hydrogen

Scientists are also designing cars that use hydrogen. Here's how it works.

Hydrogen is a gas. It has no color or smell.

1. Hydrogen goes into a fuel cell.

2. A special process gives it an electric charge.

3. The electricity powers the car.

4. The hydrogen mixes with oxygen from the air. This makes water. It is a waste product.

5. When the hydrogen is used up, the car gets refueled.

USING LESS

It takes energy to make plastic bottles. Many get used only once. But you can use a reusable water bottle over and over!

We use energy every day. But much of what we use harms the planet. Our use of coal, oil, and gas is changing the climate. Even electricity is not always clean. A lot of it comes from power plants that burn coal or gas.

HELPING OUT

Could you use less energy? You could walk or bike instead of driving. Even riding a bus uses less energy per person. Remember that making things uses energy, too. The less you use, the more energy you save.

Saving Energy at Home

One of the best places to save energy is at home. Here are some ideas.

Turn off the lights when you leave a room.

Turn off the faucet when you're not using it.

Take quick showers instead of baths.

Don't leave the fridge standing open.

Put on a sweater instead of turning the heating up.

Turn off your devices when you're not using them.

Hang clothes up to dry instead of using a tumble dryer.

QUIZ

How much have you learned about energy?
It's time to test your knowledge!

1. Where does the human body get its energy?
a. water
b. food
c. injections

2. What does a generator turn into electricity?
a. spinning motion
b. steam
c. coal

3. Where are fossil fuels found?
a. inside living things
b. in space
c. underground

4. What kind of fuel do airplanes use?
a. wind power
b. kerosene
c. biomass

The answers are on page 24.

GLOSSARY

atmosphere the blanket of gases that surrounds the Earth

biomass living things, such as plants, that can be used as a source of fuel

carbon dioxide a gas that humans and animals breathe out, which is also produced when fossil fuels are burned

climate change the gradual changing and warming of Earth's climate

current a flow of electric charge

electricity a form of energy that flows as a current, which we can use to power devices

energy the ability to do work

fossil fuels fuels such as oil, coal, or natural gas that are made from the remains of prehistoric living things

fuel a substance that can be burned to release the energy stored inside it

generator a device that converts spinning movement into electricity

power plant a factory where other forms of energy are turned into electricity

turbine a large fan-like machine that spins when air or water flows past its blades

FIND OUT MORE

Books
Biomass Energy. Robyn Hardyman, Cheriton Children's Books, 2022.

DKfindout! Energy. Emily Dodd, Dorling Kindersley, 2018.

Infographics: Renewable Energy. Alexander Lowe, Cherry Lake Publishing, 2022.

Websites
bbc.co.uk/bitesize/topics/zshp34j/articles/zntxgwx

dkfindout.com/us/science/energy/types-energy/

eia.gov/kids/

INDEX

B C
biomass 15
climate change 17, 20
crude oil 8, 9, 16, 17, 18, 20

E F G H
electricity 6. 7, 8, 10, 11, 12, 13, 14, 16, 18, 20
food 4, 14, 15, 16
fossil fuels 8, 9
generators 7, 10, 11, 12
heating 12, 14, 15, 21
homes 4, 5, 6, 13, 14, 21
hydroelectric power 11
hydrogen power 19

L P
lights 5, 6, 13, 21
power plants 6, 7, 8, 20

R S T
renewable energy 12
saving energy 20, 21
solar power 12, 13
turbines 7, 10, 11

V W
vehicles 4, 5, 16, 17, 18
wind power 10

Answers: 1. b; 2. a; 3. c; 4. b